T0225238

Cambridge Elements ≡

Elements in Geochemical Tracers in Earth System Science
edited by
Timothy Lyons
University of California
Alexandra Turchyn
University of Cambridge
Chris Reinhard
Georgia Institute of Technology

SR ISOTOPES IN SEAWATER

Stratigraphy, Paleo-Tectonics, Paleoclimate, and Paleoceanography

B. Lynn Ingram
University of California Berkeley

Donald J. DePaolo
University of California Berkeley

CAMBRIDGE
UNIVERSITY PRESS

CAMBRIDGE
UNIVERSITY PRESS

University Printing House, Cambridge CB2 8BS, United Kingdom

One Liberty Plaza, 20th Floor, New York, NY 10006, USA

477 Williamstown Road, Port Melbourne, VIC 3207, Australia

314–321, 3rd Floor, Plot 3, Splendor Forum, Jasola District Centre,
New Delhi – 110025, India

103 Penang Road, #05–06/07, Visioncrest Commercial, Singapore 238467

Cambridge University Press is part of the University of Cambridge.

It furthers the University's mission by disseminating knowledge in the pursuit of
education, learning, and research at the highest international levels of excellence.

www.cambridge.org
Information on this title: www.cambridge.org/9781108994293
DOI: 10.1017/9781108991674

© B. Lynn Ingram and Donald J. DePaolo 2022

First published 2022

A catalogue record for this publication is available from the British Library.

ISBN 978-1-108-99429-3 Paperback
ISSN 2515-7027 (online)
ISSN 2515-6454 (print)

Sr Isotopes in Seawater

Stratigraphy, Paleo-Tectonics, Paleoclimate, and Paleoceanography

Elements in Geochemical Tracers in Earth System Science

DOI: 10.1017/9781108991674
First published online: March 2022

B. Lynn Ingram
University of California Berkeley

Donald J. DePaolo
University of California Berkeley

Author for correspondence: Donald J. DePaolo, depaolo@eps.berkeley.edu

Abstract: Studies of Sr isotopic composition of thousands of samples of marine sediments and fossils have yielded a curve of $^{87}Sr/^{86}Sr$ versus age for seawater Sr that extends back to 1 billion years. The ratio has fluctuated with large amplitude during this time period, and because the ratio is always uniform in the oceans globally at any one time, it is useful as a stratigraphic correlation and age-dating tool. The ratio also appears to reflect major tectonic and climatic events in Earth history and hence provides clues as to the causes, timing, and consequences of those events. The seawater $^{87}Sr/^{86}Sr$ ratio is generally high during periods marked by continent–continent collisions and lower when continental topography is subdued, and seafloor generation rates are high. There is evidence that major shifts in the seawater ratio can be ascribed to specific orogenic events and correlated with large shifts in global climate.

Keywords: isotopes, seawater, stratigraphy, geochronology, paleoceanography

ISBNs: 9781108994293 (PB), 9781108991674 (OC)
ISSNs: 2515-7027 (online), 2515-6454 (print)

Contents

Introduction

The strontium isotope study of marine sediments and fossils shows that the $^{87}Sr/^{86}Sr$ ratio of seawater has fluctuated over the Phanerozoic Eon, which spans the past 540 million years. The most complete compilation of the available Phanerozoic data is provided in McArthur et al. (2012, 2020) and a figure adapted from that work is provided as Figure 1. The original source containing the first sufficiently comprehensive data set for this time period can be found in Burke et al. (1982), and an earlier version was published by Veizer and Compston (1974). The fluctuations in the $^{87}Sr/^{86}Sr$ ratio, which are inferred to be a globally uniform signature of seawater, can be used as stratigraphic markers to evaluate the age of marine Sr-bearing minerals. In this Element, we review the data collection that has gone into establishing this isotopic record of seawater and the assumptions, limitations, and other considerations that are involved in its use in geology. We also discuss the global geologic processes that are believed to be responsible for the fluctuations in the isotopic record. To the extent that those can be understood, the seawater $^{87}Sr/^{86}Sr$ record is a record of global Earth tectonic, erosional, and weathering processes (cf. Turchyn and DePaolo, 2019). The most recent research is focused on better understanding the modern Sr geochemical cycle as a model for understanding the variations evident in the geologic past (e.g. Mokadem et al., 2015; Peucker-Ehrenbrink and Fiske, 2019).

The rationale and process that goes into establishing the seawater $^{87}Sr/^{86}Sr$ curve, as shown in Figure 1, begins with measurements of the isotope ratio made on arguably unaltered fossil calcite and aragonite of known age (Burke et al., 1982). Generally, such samples are from sediments deposited in open marine environments far from freshwater influence and where the stratigraphy is known well and tied to geochronology by various means. This exercise has been repeated extensively enough over the past 50 years to develop a good representation of seawater $^{87}Sr/^{86}Sr$ over the Phanerozoic (McArthur et al., 2012, 2020) and a much less detailed record for the Neoproterozoic (e.g. Goddéris et al., 2017; Figure 2). The reference curve can theoretically then be used to determine the age of other carbonate-containing marine sediments that do not have useful biostratigraphy or geochronology. As shown in Figure 1, the measurement of a single sample is not likely to determine its age even if all other necessary criteria are satisfied because the curve is not monotonic except over the past 40 million years. However, in certain instances, useful age constraints could come from a single measurement. For example, if the $^{87}Sr/^{86}Sr$ ratio is lower than 0.7071, the sample would need to be either about 160 or 260 Ma. Ratios higher than 0.7085 occur only in

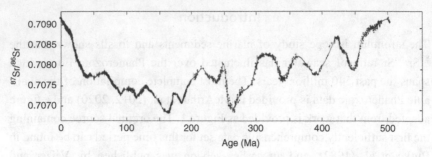

Figure 1 Consolidated seawater $^{87}Sr/^{86}Sr$ curve constructed from figures provided in McArthur et al. (2012). Due to small distortions of the age axis in this figure, the original publication should be consulted for use in Sr isotope geochronology. The data used to produce this curve were filtered by the authors in an unspecified way, but nevertheless, this representation of the seawater curve is widely regarded as the best available. An updated version of the curve is provided in McArthur et al. (2020), and a slightly higher resolution curve for the past 40 million years is provided in Figure 3

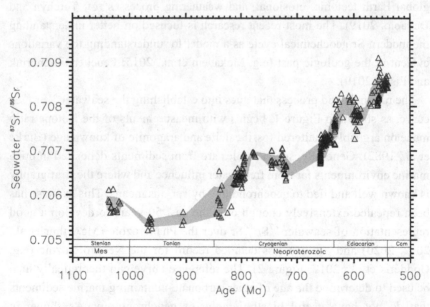

Figure 2 Sr isotope data compiled by Goddéris et al. (2017) for the Neoproterozoic and latest Mesoproterozoic. The age axis is plotted with age increasing to the left, opposite to what is shown in the other figures. The gray-shaded area represents the authors' estimates of 1 Σ uncertainty on the $^{87}Sr/^{86}Sr$ ratio. Park et al. (2020) provide another version for parts of this time period. The period from 800 to 540 Ma corresponds to the assembly of Gondwanaland, but other interpretations have also been proposed (cf. McArthur et al., 2020)

the past 20 million years, at 410–420 Ma, or prior to 460 Ma. If the sample has an intermediate ratio, then other information, or a more continuous record spanning a significant time period, are necessary to match the pattern and the range of $^{87}Sr/^{86}Sr$ to the curve (e.g. Korte et al., 2003; Swanson-Hysell and Macdonald, 2017).

Sources and Sinks of Sr in the Oceans

There are several sources of dissolved strontium to the oceans, including rivers draining the world's continents and islands, Sr dissolved in hydrothermal fluids emanating from mid-ocean ridges, lower temperature fluids entering the ocean from weathering of seafloor basalt, diagenesis of ocean floor sediments, and continental groundwaters entering the oceans in coastal areas of both continents and ocean islands (Veizer, 1989; Beck et al., 2013). In detail, the relative proportions coming from each source are still a matter of debate (Allegre et al., 2010; Peucker-Ehrenbrink and Fiske, 2019), as is the question of whether the sources and sinks are balanced in the modern oceans. Strontium is removed from the oceans mainly by biogenically precipitated calcium carbonate, with Sr partially replacing calcium in the calcite or aragonite crystal structure. Strontium is also removed by the inorganic precipitation of chemical sediments such as evaporites, ferromanganese nodules, and crusts. In general, the removal flux of Sr from the oceans is only slightly different from the input flux, causing the average concentration of Sr in the oceans to shift slowly over millions of years (Steuber and Veizer, 2002). The ratio Sr/Ca in the oceans has changed very little over the past 200 million years (e.g. Lear et al., 2003; Turchyn and DePaolo, 2019), but the concentration of Ca in the oceans has been as much as three times higher than the modern value of about 10 mM (Lowenstein et al., 2001). It is therefore inferred that the seawater Sr concentration has also varied by about this factor. The modern oceans have a concentration of about 8 ppm Sr (90 µM), but the concentration may have been three times larger at 90–100 Ma (Steuber and Veizer, 2002), and if Ca concentration is a reliable guide, some estimates for the Precambrian are much higher still (Halevy and Bachan, 2017).

The element Sr has four isotopes: 84, 86, 87, and 88. All of these isotopes are stable, but one is radiogenic (strontium-87 or ^{87}Sr). Strontium-87 is produced by the radioactive decay of rubidium-87. Because continental rocks are relatively old and have a much higher concentration of rubidium than oceanic rocks (high $^{87}Rb/^{86}Sr$), they have a significantly higher proportion of ^{87}Sr and a higher $^{87}Sr/^{86}Sr$ ratio than rocks on the seafloor. Oceanic crust is geologically young and relatively depleted in rubidium (low $^{87}Rb/^{86}Sr$) and has a lower proportion

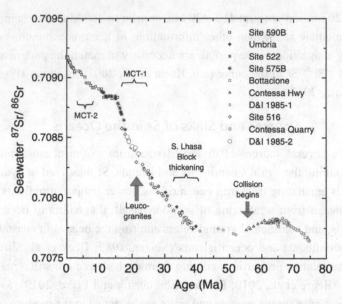

Figure 3 Data for the past 75 million years mostly produced at University of California Berkeley, all normalized to a modern seawater $^{87}Sr/^{86}Sr = 0.709175$. These data show a more detailed structure in the curve for 0–40 Ma and include diagenesis corrections to data from sites 575B and 590B. Comparison of this curve to that in Figure 1 provides an indication of the degree of smoothing incorporated into the MacArthur et al. (2020) compilation. (Bottacione and Contessa Highway data from Sinnesael et al., 2019; D&I, Site 522, and Contessa Quarry from DePaolo and Ingram, 1985; sites 575B and 590B from DePaolo, 1986, and Richter and DePaolo, 1988; Site 516 from Hess et al., 1986; and Umbria are unpublished data of the authors.) Time of thickening of the southern Lhasa Block is from DePaolo et al. (2019). MCT-1 and MCT-2 are periods of major movement on the Main Central Thrust underlying the Himalayas. Evidence of rapid erosion of the Himalayas between 20 and 16 Ma as recorded in the leucogranites is described in Chen et al. (1996)

of ^{87}Sr. Measurements of seafloor basalts and hydrothermal vents yield a typical $^{87}Sr/^{86}Sr$ ratio of 0.703, while the global average $^{87}Sr/^{86}Sr$ ratio in river waters draining the continents is 0.711 (Veizer, 1989; Peucker-Ehrenbrink and Fiske, 2019).

In the modern ocean, approximately 75 percent of Sr input comes from continental sources (with high $^{87}Sr/^{86}Sr$) versus 25 percent from the seafloor and oceanic islands (low $^{87}Sr/^{86}Sr$). The inputs from all of the sources of strontium, including the dissolved constituents in rivers draining the continents, hydrothermal vents at mid-ocean ridges, and the weathering of the seafloor, volcanic plateaus, and volcanic arcs, are mixed and homogenized in the world ocean. Due to thermohaline circulation, the modern ocean mixes quickly in comparison to the rate at which Sr is added and removed, so the Sr concentration and $^{87}Sr/^{86}Sr$ ratio in seawater are nearly uniform throughout the ocean basins. The extent of this uniformity is discussed in the next section.

In contrast to the lighter stable isotopes like oxygen and carbon, strontium isotopes are not fractionated by biological processes or evaporation or at least not in a manner that affects the $^{87}Sr/^{86}Sr$ ratio. The $^{87}Sr/^{86}Sr$ ratio in seawater therefore exhibits no change with depth or location in the oceans, and the ratio measured in fossil carbonates and other marine precipitates reflects that of the ambient seawater during the time the organism lived. The Sr in marine biogenic precipitates does undergo a small amount of mass-dependent fractionation during the precipitation process, typically changing the $^{88}Sr/^{86}Sr$ ratio by about 0.01–0.03 percent (Müller et al., 2018). This fractionation does not affect the measured $^{87}Sr/^{86}Sr$ ratio, however, because the measurement of $^{87}Sr/^{86}Sr$ involves normalizing to a constant $^{88}Sr/^{86}Sr$ ratio to account for any mass-dependent isotopic fractionation in nature or in the mass spectrometer used for the measurement.

How Uniform Is the $^{87}Sr/^{86}Sr$ of Modern Seawater?

The application of seawater Sr isotopes as a stratigraphic tool requires that the $^{87}Sr/^{86}Sr$ ratio of seawater be the same everywhere in the oceans at all times, even though the ratio is drifting slowly up and down over millions of years. It is useful to ask how good an assumption this is. In theory, the variability of the ratio in seawater should be predicted by the ratio of the Sr residence time in the oceans (τ_{res}) to the time required to mix the ocean by stirring with ocean currents (τ_{mix}). The variability is also dependent on the range in $^{87}Sr/^{86}Sr$ of the Sr being delivered to the oceans. The Sr in the oceans comes mostly from what is delivered in dissolved form by rivers. It is estimated that the global average value for rivers is about 0.711, but the documented range of values for major rivers is 0.7095 (Mississippi) to 0.725 (Ganges) (Goldstein and Jacobsen, 1987; Allegre et al., 2010). Also, about 20–30 percent of the Sr delivered to the oceans comes from hydrothermal reactions and weathering of seafloor basalts, and oceanic islands (Allegre et al., 2010). The $^{87}Sr/^{86}Sr$ of the Sr being delivered to the oceans from these volcanic sources is about 0.703. Using these data, the total variation in $^{87}Sr/^{86}Sr$ of seawater Sr sources is $0.725 - 0.703 = 0.022$. This number multiplied by the ratio τ_{mix}/τ_{res} provides a good estimate of the expected variability of $^{87}Sr/^{86}Sr$ in seawater. The value of τ_{res} is known reasonably well for the modern oceans and is about 2.7 million years (Palmer and Edmond, 1989). The mixing time of the oceans that applies for this situation is more than 1,000 years but probably less than 10,000 years based on radiocarbon measurements and other estimates (Broecker and Peng, 1982). Given these numbers, the expected range of $^{87}Sr/^{86}Sr$ of seawater at any one time is ≤ 0.00001. This range is approximately the same as the analytical uncertainty on a single modern measurement and generally smaller than observed variability in samples from the same sedimentary formation at the same stratigraphic horizon. Consequently, the assumption of uniform seawater $^{87}Sr/^{86}Sr$ at all times is not a limitation of the method.

There have been studies aimed at determining the uniformity of Sr isotopes in the modern oceans. DePaolo and Ingram (1985) measured fifteen globally distributed modern shell samples and found a standard deviation of 0.000009, which suggests that all of the variability is statistical and due to the measurement uncertainty. Capo and DePaolo (1992) demonstrated methods for measuring the $^{87}Sr/^{86}Sr$ with a resolution of ± 0.000005 and found no differences between several seawater samples in the Atlantic and Pacific Oceans and Hudson Bay. Kuznetsov et al. (2012) measured twenty-five samples of bivalve, gastropod, and Anthozoa shells growing in seawater in the Caribbean, Northeast Atlantic, Indian, and Western Pacific Oceans and could find no differences in measured $^{87}Sr/^{86}Sr$ greater than ± 0.000003. Huang et al. (2011) looked specifically for effects due to freshwater dilution in coastal waters and were able to document small effects (mostly <0.00003) that correlate with salinity. Surface water tends to be most affected; deeper waters are homogeneous and the same as the global $^{87}Sr/^{86}Sr$ value. In more restricted estuarine waters, such as San Francisco Bay, deviations as large as –0.0002 have been measured in waters with a salinity of 7‰, which is much lower than the typical seawater salinity of 35‰ (Ingram and Sloan, 1992; Ingram and DePaolo, 1993). In the Bay of Bengal, where riverine input from the Ganges and Brahmaputra systems and groundwater input have exceptionally high $^{87}Sr/^{86}Sr$, deviations of the ratio from the average seawater values are as large as +0.0001 in water with salinity that is not much different from that of average seawater (Chakrabarti et al., 2018). Muller and Mueller (1991) showed that the isolation of the Mediterranean Sea from the global oceans due to desiccation in the latest Miocene is reflected in $^{87}Sr/^{86}Sr$ ratios displaced by up to –0.0005 in the local sediments from this time period.

Overall, the assumption of globally uniform seawater $^{87}Sr/^{86}Sr$ is excellent with only a few potential exceptions for the modern oceans wherever seawater salinity does not deviate from average seawater by more than a few per mil. In earlier parts of the Phanerozoic, the seawater Sr concentration may have been higher (Steuber and Veizer, 2002) and the residence time longer, which would probably make the variability in $^{87}Sr/^{86}Sr$ even smaller. Counteracting this effect, during warm climate conditions, deep circulation in the oceans could be more sluggish, and hence, the mixing time may be somewhat longer.

Measuring $^{87}Sr/^{86}Sr$ Ratios

Measuring $^{87}Sr/^{86}Sr$ in any seawater-derived mineral requires mainly a high enough Sr concentration that there is an adequate amount for analysis. Strontium has the same charge as calcium, a major constituent of seawater (both are 2+ cations), but is a slightly larger ion. Sr can be substituted into calcite, for example, but it tends to be partially excluded so that typical biogenic calcite has Sr/Ca that is about 0.1–0.2 seawater ratio (e.g. Lear et al., 2003). This ratio does not represent equilibrium between calcite and seawater, but rather is kinetically controlled because the organisms produce shell material from over-saturated solutions and the precipitation rates are quite high. The equilibrium calcite Sr/Ca is expected to be only about 0.020–0.025 that of seawater at 5 to 25°C (Zhang and DePaolo, 2020).

The majority of Sr isotope studies utilize carbonate fossils, which are composed of the minerals calcite (foraminifers, coccoliths, and macrofossils) and aragonite (pteropods, corals, and various macrofossils). Sr concentrations in carbonate minerals vary over a large range. Pre-Cenozoic limestones and their included fossils tend to have relatively low concentrations of 100–500 ppm. Cenozoic age coccolithophores typically have 1,000–2,000 ppm Sr. Foraminiferal Sr concentrations are 800–1,500 ppm. Aragonite typically has much higher concentrations of Sr in the 5,000–10,000 ppm range because the crystal structure of aragonite is more suited to incorporation of Sr substituted for Ca. Inorganic mineral precipitates from seawater, such as evaporites, ferromanganese nodules, and encrustations, can have enough incorporated Sr to measure (Ingram et al., 1990). In Paleozoic and Mesozoic limestones, it has been found that conodonts (Ca-phosphate) can be a reliable recorder of seawater $^{87}Sr/^{86}Sr$ (e.g. Korte et al., 2003).

Mass spectrometric methods continue to improve, but currently, high precision measurements of $^{87}Sr/^{86}Sr$ (± 0.00001) can be made on a sample of 100 ng of separated pure Sr and with care on smaller samples down to less than 10 ng. In a fossil calcite sample having 1,000 ppm Sr, 100 ng is the amount contained in 100 μg of calcite. A typical individual foraminifer test has a mass of a few micrograms, so precise $^{87}Sr/^{86}Sr$ measurements require multiple foraminifer tests. In larger shells such as gastropods or bivalves, subsampling of shells is possible to track changes in the local seawater $^{87}Sr/^{86}Sr$ during the growth of the shell. In coastal waters, this type of information can yield information on a local river or groundwater discharge.

For a typical mass spectrometric measurement of $^{87}Sr/^{86}Sr$ in calcite, the calcite is first dissolved, and then purified Sr is obtained using ion-exchange chromatography to separate Sr from Ca, Mg, and other elements in the shell material. The separated Sr, contained in a solution of a weak acid (typically HCl or HNO_3) is deposited on small rhenium or tantalum filament, dried, and then placed under vacuum in a thermal ionization mass spectrometer and heated to about 1,200°C by passing a current of 2 A or more through the filament. Standard mass spectrometric analysis methods allow for measurement of the relative ion beam intensities at masses 84, 86, 87, and 88, and these intensities are used to derive the ratios of $^{87}Sr/^{86}Sr$ and $^{86}Sr/^{88}Sr$ in the ion beam. The ion beam isotope intensity ratios are generally fractionated (i.e. systematically shifted as a function of atomic mass) relative to the isotope ratios in the Sr sample on the filament, and the measured ratios gradually drift during the measurement. The ion beam intensity ratios are typically sampled at 100 separate times during measurement. The $^{87}Sr/^{86}Sr$ ratio is calculated by determining how different the measured $^{86}Sr/^{88}Sr$ ratio is from the accepted value of 0.1194 and then using this difference to correct the $^{87}Sr/^{86}Sr$ ratio to the value that corresponds to $^{86}Sr/^{88}Sr = 0.1194$ for each of the 100 sampling times. The corrected $^{87}Sr/^{86}Sr$ ratios are averaged in groups of ten, and the average $^{87}Sr/^{86}Sr$ is then calculated as the standard deviation of the mean of the averages of the ten groups of ten measurements. The 95 percent uncertainty, or $2_{\sigma mean}$, is reported as the analytical uncertainty and is typically ±0.00001 or less. In order to allow different laboratories to compare their results, $^{87}Sr/^{86}Sr$ ratios are also normalized either to either or both of a modern carbonate standard that represents the seawater $^{87}Sr/^{86}Sr$ ratio or a more generally used isotopic standard. Most laboratories use the US Geological Survey Sr isotope standard; EN-1 (a modern Tridacna clamshell from the Eniwetok Lagoon), which has $^{87}Sr/^{86}Sr$ equal to the average modern seawater value; and the National Bureau of Standards SRM-987 for interlaboratory comparison.

In recent years, $^{87}Sr/^{86}Sr$ ratios are also being measured using inductively coupled plasma mass spectrometry. The procedures are similar to those described previously except that the Sr is introduced to the mass spectrometer as a dilute acid solution and no filaments are involved. Measurement precision can be as good as with thermal ionization, but reproducibility can be somewhat poorer, and matrix effects still require a fairly pure Sr solution so chemical separation of Sr from Ca and other elements is still desirable.

Dating and Correlating Marine Sediments Using Sr Isotopes

The early decades of Sr isotopic research on marine sediments focused on constructing a marine Sr isotope curve to be used for dating and correlating marine sedimentary rocks. These studies focused on ^{87}Sr/^{86}Sr measurements of well-dated and well-preserved carbonate fossils. Now established, this strontium isotope seawater curve has been used to date marine carbonates of unknown age. Over the past two decades, statistical regression functions have been developed to allow researchers to easily convert ^{87}Sr/^{86}Sr ratios in samples of unknown age to numerical ages (McArthur et al., 2001).

Using strontium isotopes for dating and correlating marine carbonates assumes the marine organisms grew in fully marine conditions as noted previously. Marginal marine or coastal settings in close proximity to the mouths of rivers, such as estuaries and bays, typically have brackish water conditions due to dilution with freshwater. The ^{87}Sr/^{86}Sr ratios of these waters, and therefore any carbonates precipitated in them, can be different from fully marine seawater, but these differences can be used as a measure of paleosalinity variations, which in turn can relate to paleo-river discharge and paleo-precipitation (Ingram and Sloan, 1992).

Biogenic carbonates are produced in most surface waters, and after organisms die, their remains sink to the bottom where they become part of the sedimentary record. Because carbonate minerals begin to dissolve in seawater with increasing water depth, sediments deposited deeper than the level at which carbonate begins to dissolve rapidly (the carbonate compensation depth), often lack material suitable for Sr isotopic measurements. These deep-sea clays and siliceous deposits sometimes contain the skeletal remains of fish (composed of the mineral apatite) that contain strontium that can be measured for ^{87}Sr/^{86}Sr (e.g. Ingram, 1995).

Once deposited on the seafloor, strontium in the carbonate sediments can exchange with that in the trapped pore fluids surrounding them, causing the original ^{87}Sr/^{86}Sr ratios of the carbonates to shift toward the ratio of these pore fluids. As sediments are compacted and lithify over time, they continue to recrystallize and undergo diagenesis. Models have been developed to estimate how much change in the ^{87}Sr/^{86}Sr ratio might accrue in bulk carbonate sediment from post-depositional diagenesis (Richter and DePaolo, 1988; Fantle and DePaolo, 2006). These models apply to the earlier stages of diagenesis when the carbonate sediments are still oozes, or chalks, but become more complicated and less reliable as the sediments are lithified. The models also do not yield information on any specific component of the sediments, so most practitioners are more interested in identifying and sampling specific materials that appear, based on other criteria, to be unaffected by diagenesis. In general, to determine a reliable Sr isotopic value, samples must be chosen carefully and assessed by several means to minimize the likelihood they have undergone significant diagenetic alteration. Older marine fossils exposed in sedimentary sections on continents are a particular problem, as most of these materials have more complex histories of lithification, structural deformation, and potential alteration by groundwaters after they have been uplifted or otherwise exposed on land.

In the original work on the Phanerozoic seawater $^{87}Sr/^{86}Sr$ curve by Burke et al. (1982), the authors highlighted the need for particular attention to be directed at alteration (or diagenesis) of fossil materials over geologic time, which may result in shifts in the original $^{87}Sr/^{86}Sr$ ratio of these materials. Their curve illustrated that with increasing age, there is an increasing range in the $^{87}Sr/^{86}Sr$ of the carbonate samples they chose as being representative of seawater. Burke et al. (1982) attributed this scatter to the reaction of the carbonates with high $^{87}Sr/^{86}Sr$ groundwaters. Subsequently, it has been generally assumed that post-depositional alteration will shift the $^{87}Sr/^{86}Sr$ ratio of carbonates to higher values. A good example of the application of this principle is shown in the work of Shields and Veizer (2002), which is an attempt to construct a seawater $^{87}Sr/^{86}Sr$ curve for the Precambrian, where it appears that almost all of the samples are altered, and some to a large degree. A demonstration of the degree to which the seawater Sr isotope curve can be established for the latest Precambrian (1,000–540 Ma; Figure 2) is the data assembled by Goddéris et al (2017), which clearly show a trend of increasing $^{87}Sr/^{86}Sr$ starting at about 860 Ma, but the age resolution afforded by the $^{87}Sr/^{86}Sr$ ratios is much poorer than that from radiometric dating methods (e.g. Park et al., 2020).

The Geologic Meaning of Seawater $^{87}Sr/^{86}Sr$ Fluctuations

The changes in the $^{87}Sr/^{86}Sr$ ratio of seawater over geologic time (Figures 1 and 2) are assumed to contain information about past global-scale tectonic processes, ocean chemistry, and climate. The prevailing view is that the $^{87}Sr/^{86}Sr$ ratio of seawater reflects the relative proportion of high-$^{87}Sr/^{86}Sr$ strontium entering the oceans from the continents versus low-$^{87}Sr/^{86}Sr$ strontium coming from the seafloor and ocean island weathering. This model implies that during times of intense mountain building due to continent–continent collisions, the seawater $^{87}Sr/^{86}Sr$ should increase or remain high, and during times when continents are breaking apart and there are few collisions, the seawater $^{87}Sr/^{86}Sr$ should decrease or remain low (Turchyn and DePaolo, 2019). Low $^{87}Sr/^{86}Sr$ could also be a consequence of higher global seafloor generation rates, or higher rates of oceanic magmatism like island arcs, submarine plateaus, and other hot spot manifestations. The increased spreading rates and oceanic volcanism might be correlated, and should also be correlated, with high sea levels, warm climates, and anoxic oceans. This relatively simple model may be largely correct but is shown to be incomplete by some recent data, particularly where mountain building involves uplift and erosion of rocks from the ocean floor or island arcs (e.g. Swanson-Hysell and Macdonald, 2017).

Late Cenozoic Rise in $^{87}Sr/^{86}Sr$

The Late Cenozoic is a useful time period to analyze both the origins of seawater Sr isotope variations and the prospects for using the seawater curve for detailed stratigraphy. It is evident from Figure 1 that seawater $^{87}Sr/^{86}Sr$ has risen monotonically over the past 40 million years and that there were also significant increases during the Late Cretaceous and the Late Jurassic–Early Cretaceous time periods. The Sr data compilation for the past 75 million years shown in Figure 3, taken from the work of the authors and colleagues, shows in detail the rise in $^{87}Sr/^{86}Sr$ since 40 Ma, a period where there is a good possibility of associating the Sr isotopes with tectonic events, as well as changes in ocean chemistry and global climate. This data set, including corrections for diagenesis in data from sites 590B and 575 (Richter and DePaolo, 1988), shows the high level of coherence that can be expected and the resulting stratigraphic or age resolution.

As shown in Figure 3, it is possible to associate the rapidly rising seawater $^{87}Sr/^{86}Sr$ with India – Asia continental collision and the formation of the Himalaya. The connection was first proposed by Raymo et al. (1988), who argued that Sr isotopes indicated a large increase in silicate weathering of continental rocks high in $^{87}Sr/^{86}Sr$, and this increase in continental silicate weathering could have resulted in the removal of atmospheric CO_2 and hence, ultimately, a cooling of global climate. The timing seems appropriate, as the Antarctic icecap began to form and grow at about 35 Ma, and is likely to have been associated with a drop in atmospheric CO_2 below 560–700 ppm (DeConto and Pollard, 2003). Richter et al. (1992) showed that the periods of the most rapid erosion of the Himalayas and Tibet were closely correlated with periods of the most rapid increase in the $^{87}Sr/^{86}Sr$ ratio of seawater. In addition, the total amount of sedimentary material eroded off the uplifted regions is appropriate for the amount of additional erosion necessary to account for the overall increase in seawater $^{87}Sr/^{86}Sr$. The analysis of Richter et al. (1992) concluded that both the flux of strontium dissolved in rivers as well as the $^{87}Sr/^{86}Sr$ ratio of rivers increased, and both were primarily a function of mountain building and associated increased erosion related to the India–Asia collision. A key implication of this work is that significant excursions in the seawater Sr isotope curve can be attributed to a single large orogen.

The rise in seawater $^{87}Sr/^{86}Sr$ paused between 15 and 12 Ma and then resumed at a high rate to the present (Capo and DePaolo, 1990). The latter part of this period corresponds to when the northern hemisphere ice caps came into being. Although it has been argued that it is unlikely that the Sr isotope curve should reflect major changes in weathering fluxes and hence atmospheric CO_2 drawdown (e.g. McCauley and DePaolo, 1997), there is nevertheless a correlation between the steep $^{87}Sr/^{86}Sr$ increases in the Late Cenozoic and the development of polar ice caps. The primary effect may come from the fact the Himalayas and adjoining regions have high erosion rates *and* exceptionally high-$^{87}Sr/^{86}Sr$ rocks exposed to erosion. This signature might be applicable to other orogenic events if those events also involve uplift and rapid erosion of old, high-Rb continental rocks.

Mesozoic Oceanic Anoxic Events

The Mesozoic was dramatically different from the Late Cenozoic: the climate was warmer, there were no ice caps, the sea level was high, and the global seafloor generation rate may have been up to 30–50 percent higher (Engebretson et al., 1992). The higher volcanic eruption rates, coupled with subdued continental topography due to an absence of continent–continent collisions, may have led to significantly higher atmospheric carbon dioxide levels and warmer oceanic and atmospheric temperatures (Turchyn and DePaolo, 2019). Lower thermal gradients between the poles and equator led to sluggish ocean circulation. These conditions are thought to have led to discrete periods of extreme oxygen depletion in seawater, the so-called oceanic anoxic events (OAEs). In the geologic record, OAEs are represented by black shale formations with unusually high organic carbon content (up to 30 percent), increased $^{13}C/^{12}C$ in the oceans, lack of carbonate fossils, and abundant pyrite. Each of the black shale occurrences is global, spans an interval of one million years or less, and is thought to require anoxic bottom waters. An estimated 60 percent of the world's oil may have originated during these events (Arthur and Schlanger, 1979). The times of the recognized OAEs (Jenkyns, 2010) are indicated in Figure 4.

The several hypotheses that have been put forth to explain the origin of OAEs are reviewed in Jenkyns (2010). The OAEs may have been produced by an increase in biological productivity through enhanced nutrient inputs and enhanced organic carbon preservation because of decreased oxygen levels in bottom waters. They also may have resulted from increased seafloor volcanism, which would have raised global sea level, expanding the preservation of carbonates on continental margins. Higher rates of submarine volcanism

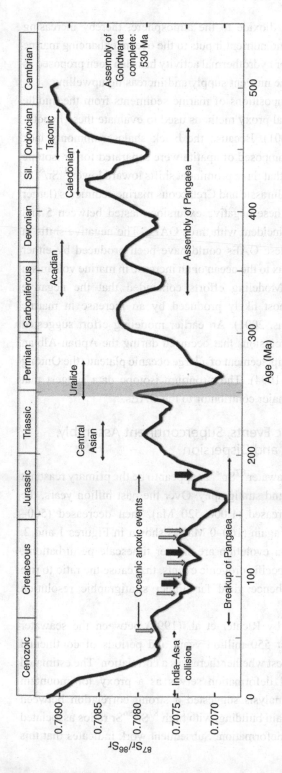

Figure 4 Annotated Phanerozoic seawater $^{87}Sr/^{86}Sr$ versus time curve. Time periods of the assembly and breakup of Pangaea are indicated, the former associated with generally higher $^{87}Sr/^{86}Sr$ and the latter with lower $^{87}Sr/^{86}Sr$. Also shown are the times of recognized ocean anoxic events (OAE) and the approximate time periods for several large orogenic events. The darker arrows are the major OAE's as described in Jenkyns (2010). Assembly of Gondwanaland occurred between 800 and 530 Ma and corresponds to the largest extended increase in seawater $^{87}Sr/^{86}Sr$ in Earth history (Figure 2)

would also add more carbon dioxide to the atmosphere, thereby increasing continental weathering rates and nutrient inputs to the ocean, enhancing marine productivity. Increased seafloor hydrothermal activity has also been proposed as a mechanism for increasing the nutrient supply and increasing upwelling.

The strontium isotopic compositions of marine sediments from the middle Cretaceous were one of several proxy methods used to evaluate these hypotheses (Jones and Jenkyns, 2001). Because the black shales commonly lack carbonate fossils, fish teeth composed of apatite were separated for Sr isotope analyses. The results indicate that three prominent shifts toward lower $^{87}Sr/^{86}Sr$ ratios in seawater occurred in Jurassic and Cretaceous marine sediments (larger black arrows in Figure 4). These negative excursions lasted between 5 and 13 million years and were coincident with three OAEs. The negative shifts in the $^{87}Sr/^{86}Sr$ ratios during these OAEs could have been produced by either a decrease in the riverine Sr flux to the ocean or an increase in marine volcanism and hydrothermal activity. Modeling efforts concluded that the negative $^{87}Sr/^{86}Sr$ excursions were most likely produced by an increase in marine volcanism (Jones and Jenkyns, 2001). An earlier modeling effort suggested that the negative $^{87}Sr/^{86}Sr$ excursions that occurred during the Aptian-Albian were likely the result of the emplacement of a large oceanic plateau (the Ontong Java Plateau; Ingram et al., 1994). The strontium isotope data suggest that submarine volcanism was a major contributor to the OAEs.

Collisional Orogenic Events, Supercontinent Assembly, and Dispersion

The effects of tectonism on seawater $^{87}Sr/^{86}Sr$ appear to be the primary reason it is useful for geochronology and stratigraphy. Over the past billion years, the ratio has first gradually increased (1,000–520 Ma), then decreased (500–160 Ma), and then increased again (160–0 Ma) as shown in Figures 1 and 2. Superimposed on this gradual evolution are shorter timescale perturbations, presumably associated with specific orogenic events, that cause the ratio to rise and fall more steeply and hence yield far higher stratigraphic resolution (Figures 3 and 4).

A comparison was made by Richter et al. (1992) between the seawater $^{87}Sr/^{86}Sr$ curve over the past 550 million years and periods of continental contractional deformation to test whether there was a correlation. The estimated areal extent of contractional deformation serves as a proxy for mountain building and erosion. The analysis suggested a strong correlation between seawater $^{87}Sr/^{86}Sr$ and mountain building, with high $^{87}Sr/^{86}Sr$ ratios associated with increased contractional deformation. Subsequent work indicates that this view, although generally correct, may be too simple in that arc–continent collisions might have the opposite effects on seawater $^{87}Sr/^{86}Sr$ than continent–continent collisions.

Swanson-Hysell and Macdonald (2017) analyzed a 35-million-year cooling trend through the Ordovician, culminating with glaciation and mass extinction in the end Ordovician. Several hypotheses had been proposed for the cause of this transition from ice-free to glaciated conditions, including reduced volcanic outgassing, increased carbon burial, increased weathering of volcanic rock, and increased silicate weathering associated with a major orogeny (the Taconic arc–continent collision). Detailed geochronology and comparison to the Ordovician Sr isotope curve were used as a test of the Taconic orogeny hypothesis. The seawater Sr curve for this time period is based on measurements on fossil conodonts (apatite) and brachiopod calcite (Saltzman et al., 2014). A paleogeographic reconstruction of the Ordovician demonstrates that the paleocontinent Laurentia was close to the equator at 465 Ma, in the wet and warm tropics where chemical weathering rates are highest. The collision and exhumation of the Taconic arc system, which is marked by the appearance of detrital chromite in foreland basins, correlate with a decrease in the seawater $^{87}Sr/^{86}Sr$ ratio and hence are consistent with tropical weathering of the Taconic arc–continent collision as a driver of Ordovician cooling. The $^{87}Sr/^{86}Sr$ ratio decreased from 0.7090 to 0.7088 between 480 and 465 million years ago and decreased more steeply between 465 and 450 Ma to 0.7079.

The sense of the shift in seawater $^{87}Sr/^{86}Sr$ for this example is opposite to that of the India–Asia collision because the Taconic arc–continent collision resulted in uplift and exposure to rapid weathering of mafic and ultramafic rocks with low $^{87}Sr/^{86}Sr$. The chemical weathering of these rocks would nevertheless have sequestered CO_2 quite efficiently, thereby causing a decrease in global temperature throughout the Ordovician leading to glaciation, much like the India–Asia collision has apparently done in the Late Cenozoic.

The period straddling the Permo–Triassic boundary at 250 Ma is one with the most rapid increase in seawater $^{87}Sr/^{86}Sr$ in the Phanerozoic (MacArthur et al., 2020). This event has received relatively little attention in the literature until recently (Sedlacek et al., 2014; Song et al., 2015). The prevailing view is that the steep increase reflects an increased flux of high-$^{87}Sr/^{86}Sr$ strontium from continental weathering, but there is no mention of any specific orogenic event. There was a significant continent–continent collision at this time that resulted in the formation of the Ural Mountains (Zuza and Yin, 2017), although it is unclear whether the timing is right. Based on the India–Asia example, it is possible that a single major orogen could be responsible for this $^{87}Sr/^{86}Sr$ increase, especially since this collision involved exposure of continental rocks that were old and presumably high in $^{87}Sr/^{86}Sr$. In this instance, independent of the cause of the increase, the rate of increase of $^{87}Sr/^{86}Sr$ could have been enhanced because both the concentration of Sr in seawater as well as the $^{87}Sr/^{86}Sr$ ratio of seawater 250 million years ago were lower than at almost any other time in the Phanerozoic (Steuber and Veizer, 2002). This combination means that the effect of an increased high-$^{87}Sr/^{86}Sr$ weathering flux from an orogen would have been substantially enhanced.

Concluding Remarks

Over the past 40 years, the database for the seawater $^{87}Sr/^{86}Sr$ history, its applications for geochronology and stratigraphy, and its role as a paleotectonic and paleo-climatologic tracer have advanced substantially. The isotope ratios can be measured with very high precision, and several studies have been able to retrieve highly systematic and detailed stratigraphic progressions at time periods throughout the past 1,000 million years. There is still debate about what the excursions in the ratio mean about Earth processes, but the approaches to identifying the causes are becoming more quantitative and benefit from an ever-improving knowledge base of global processes and continuing advances in geochronology that allow contemporaneity to be evaluated far back in the geologic record. The Sr isotope history of seawater has continued to help solve problems and to offer tantalizing clues about seminal events in Earth history.

Key Papers

Burke, W. H., Denison, R. E., Hetherington, E. A. et al., 1982, Variation of seawater $^{87}Sr/^{86}Sr$ throughout Phanerozoic time. Geology, vol. 10, 516–519 (The first stratigraphically useful comprehensive Phanerozoic seawater Sr isotope curve.)

McArthur, J. M., Howarth, R. J., and Shields, G. A., 2012, Strontium isotope stratigraphy. In Gradstein, F. M., Ogg, J. G., Schmitz, M., and Ogg, G. (eds), The Geologic Time Scale 2012. Elsevier, pp. 127–144 (Comprehensive Phanerozoic seawater Sr curve showing selected data that define the curve. Update in 2020 listed in references.)

DePaolo, D. J. and Ingram, B. L., 1985, High resolution stratigraphy with strontium isotopes. Science, vol. 227, 938–941 (First demonstration that carbonate $^{87}Sr/^{86}Sr$ ratios could be reliably measured to ±0.000025 and that the Cenozoic seawater curve could potentially be used for age assignments at a level almost comparable to that from biostratigraphy. Presents data from modern marine carbonate shells from different oceans that have a standard deviation of only ±0.00001.)

DePaolo, D. J., 1986, Detailed record of the Neogene Sr isotopic evolution of seawater from DSDP Site 590B. Geology, vol. 14, 103–106 (First attempt to create a high-resolution Late Cenozoic Sr isotope curve from data derived from a single deep-sea carbonate core, and showing measurement resolution, and data coherence at the ±0.00001 level or better.)

Richter, F. M., Rowley, D. B., and DePaolo, D. J., 1992, Sr isotope evolution of sea water: The role of tectonics. Earth Planet. Sci. Lett., vol. 109, 11–23 (Relates the tectonics associated with the India–Asia collision to the post-40 Ma rise in seawater $^{87}Sr/^{86}Sr$. Presents arguments that the Phanerozoic seawater curve largely represents a record of continental collision tectonics.)

Palmer, M. R. and Edmond, J. M., 1989, The strontium isotope budget of the modern ocean. Earth Planet. Sci. Lett., vol. 92, 11–26 (An early effort to systematically assess the fluxes of Sr to the oceans from oceanic and continental sources.)

Peucker-Ehrenbrink, B. and Fiske, G. J., 2019, A continental perspective of the seawater $^{87}Sr/^{86}Sr$ record: A review. Chem. Geol., vol. 510, 140–165 (Extensive data compilation providing the first quantitative estimate of the role of varying continental geology on seawater Sr isotope evolution.)

Kuznetsov, A. B., Semikhatov, M. A., and Gorokhov, I. M., 2012, The Sr isotope composition of the world ocean, marginal and inland seas:

Implications for Sr isotope stratigraphy. Stratigr. Geol. Correl., vol. 20, 501–515. © Pleiades Publishing, Ltd., 2012 (Extensive data set showing that modern carbonate shells from multiple oceans are uniform in $^{87}Sr/^{86}Sr$ to better than ±0.000005..

Allegre, C. J., Louvat, P., Gaillardet, J. et al., 2010, The fundamental role of island arc weathering in the Sr isotope budget. Earth Planet. Sci. Lett., vol. 292, 51–56 (Important summary of riverine Sr inputs to the oceans and arguing that weathering on oceanic islands constitutes a major fraction of the Sr derived from "ocean floor" type sources with low $^{87}Sr/^{86}Sr$.)

Goddéris, Y., Hir, G. L., Macouin, M. et al.,2017, Paleogeographic forcing of the strontium isotopic cycle in the Neoproterozoic. Gondwana Res., vol. 42, 151–162. http://doi.org/10.1016/j.gr.2016.09.013 (A unique effort to account quantitatively for the changing $^{87}Sr/^{86}Sr$ ratio of seawater in the Neoproterozoic using known continent configurations and known tectonic and magmatic events.)

References

Allegre, C. J., Louvat, P., Gaillardet, J. et al., 2010, The fundamental role of island arc weathering in the Sr isotope budget. Earth Planet. Sci. Lett., vol. 292, 51–56 (Important summary of riverine Sr inputs to the oceans and arguing that weathering on oceanic islands constitutes a major fraction of the Sr derived from "ocean floor" type sources with low 87Sr/86Sr.)

Arthur, M. A., and Schlanger, S. O., 1979, Cretaceous "oceanic anoxic events" as causal factors in development of reef-reservoired giant oil fields. AAPG Bull., vol. 63, 870–885.

Beck, A. J., Charette, M. A., Cochran, J. K., Gonneea, M. E., and Peiucker-Ehrenbrink, B., 2013, Dissolved strontium in the subterranean estuary – Implications for the marine strontium isotope budget. Geochim. Cosmochim. Acta, vol. 117, 33–52.

Broecker, W. S. and Peng, T. H. (1982) *Tracers in the Sea*. Eldigio Press, New York, 690pp.

Capo, R. C., and DePaolo, D. J., 1990, Seawater strontium isotopic variations: 2.5 Ma to the present. Science, vol. 249, 51–55.

Capo, R. C., and DePaolo, D. J., 1992, Homogeneity of Sr isotopes in the ocean. EOS, Trans. Am. Geophys. Union 73, 272.

Chakrabarti, R., Mondal, S., Shankar Achary, S., Lekha, J. S., and Sengupta, D., 2018, Submarine groundwater discharge derived strontium from the Bengal Basin traced in Bay of Bengal water samples. Nat. Sci. Rep., vol. 8, 4383. http://doi.org/10.1038/s41598-018-22299-5

Chen, C.-H, DePaolo, D. J., and Lan, C.-Y., 1996, Rb-Sr microchrons in the Manaslu granite: Implications for Himalayan thermochronology. Earth Planet. Sci. Lett., vol. 143, 125–135.

DeConto, R. M., and Pollard, D., 2003, Rapid Cenozoic glaciation of Antarctica induced by declining atmospheric CO_2. Nature, vol. 421, 245–249.

DePaolo, D. J. and Ingram, B. L., 1985, High resolution stratigraphy with Strontium isotopes: Science vol. 227, 938–941.

DePaolo, D. J., and Finger, K. L., 1991, High resolution strontium isotope stratigraphy and biostratigraphy of the Miocene Monterey Formation, central California.Geol. Soc. Am. Bull., vol. 103, 112–124.

DePaolo, D. J., Harrison, T. M., Wielicki, M. et al., 2019, Geochemical evidence for thin syn-collision crust and major crustal thickening between 45 and 32 Ma at the southern margin of Tibet. Gondwana Res. Vol. 73, 123–135. http://doi.org/10.1016/j.gr.2019.03.011

Engebretson, D. C., Kelley, K. P., Cashman, H. J., and Richards, M. A., 1992, 180 million years of subduction. Geol. Soc. Am. Today, vol. 2, 93–95.

Fantle, M. S., and DePaolo, D. J., 2006, Sr isotopes and pore fluid chemistry in carbonate sediment of the Ontong Java Plateau: Calcite recrystallization rates and evidence for a rapid rise in seawater Mg over the last 10 million years. Geochim. Cosmochim. Acta, vol. 70, 3883–3904.

Goldstein, S. J., and Jacobsen, J., 1987, The Nd and Sr isotopic systematics of river-water dissolved material: Implications for the sources of Nd and Sr in seawater. Chem. Geol., vol. 66, 245–272.

Gradstein, F. M., Ogg, J. G., Schmitz, M. D., and Ogg, G. M. (eds), 2020, The Geologic Time Scale. Elsevier.

Halevy, I., and Bachan, A., 2017, The geologic history of seawater pH. Science, vol. 355, 1069–1071.

Hess, J., Bender, M. L., and Schilling, J.-G., 1986, Evolution of the ratio of strontium-87 to strontium-86 in seawater from Cretaceous to present. Science, vol. 231, 979–984.

Huang, K. F., You, C. F., Chung, C. H., and Lin, I. T., 2011, Nonhomogeneous seawater Sr isotopic composition in the coastal oceans: A novel tool for tracing water masses and submarine groundwater discharge. Geochem. Geophys. Geosyst., vol. 12, 5. http://doi.org/10.1029/2010GC003372

Ingram, B. L., 1995, High-resolution dating of deep-sea clays using Sr isotopes in fossil fish teeth. Earth Planet. Sci. Lett., vol. 134, 545–555.

Ingram, B. L., and DePaolo, D. J., 1993, A 4,500-year strontium-isotope record of paleosalinity and freshwater inflow in San Francisco Bay, California. Earth Planet. Sci. Lett., vol. 119, 103–119.

Ingram, B. L., Hein, J. R., and Farmer, G. L., 1990, Age determinations and growth rates of Pacific ferromanganese deposits using Sr isotopes. Geochim. Cosmochim. Acta, vol. 54, 1709–1721.

Ingram, B. L., and Sloan, D., 1992, Strontium isotopic composition in estuarine sediments as paleosalinity and paleoclimate indicator. Science, vol. 255, 68–72.

Ingram, B. L., Coccioni, R., Montanari, A. and Richter, F. M., 1994, Strontium isotopic composition of mid-Cretaceous seawater, Science, vol. 264, 546–550.

Jacobsen, S. B., and Kaufman, A. J., 1999, The Sr, C and O isotopic evolution of Neoproterozoic seawater. Chem. Geol., vol. 161, 37–57.

Jenkyns, H. C., 2010, Geochemistry of oceanic anoxic events. Geochem. Geophys. Geosyst., vol. 11, Q03004. http://doi.org/10.1029/2009GC002788

Jones, C. E. and Jenkyns, H. C., 2001, Seawater strontium isotopes, oceanic anoxic events, and seafloor hydrothermal activity in the Jurassic and Cretaceous. Am. J. Sci., vol. 301, 112–149.

Korte, C., Kozur, H. W., Bruckschen, P., and Veizer, J., 2003, Strontium isotope evolution of Late Permian and Triassic seawater. Geochim. Cosmochim. Acta, vol. 67, 47–62. http://doi.org/10.1016 /S0016-7037(02)01035–9

Kump, L. R., 2008, The role of seafloor hydrothermal systems in the evolution of seawater composition during the Phanerozoic. In Lowell, R. P., Seewald, J. S., Metaxas, A., and Perfit, M. (eds), Magma to Microbe: Modeling Hydrothermal Processes at Ocean Spreading Centers, Geophys. Monogr. Ser., vol. 178. American Geophysical Union, pp. 275–283.

Kuznetsov, A. B., Semikhatov, M. A., and Gorokhov, I. M., 2012, The Sr isotope composition of the world ocean, marginal and inland seas: Implications for Sr isotope stratigraphy. Stratigr. Geol. Correl., vol. 20, 501–515. © Pleiades Publishing, Ltd., 2012 (Extensive data set showing that modern carbonate shells from multiple oceans are uniform in 87Sr/ 86Sr to better than ±0.000005.)

Lear, C. H., Elderfield, H., and Wilson, P. A., 2003, A Cenozoic seawater Sr/Ca record from benthic foraminiferal calcite and its application in determining global weathering fluxes. Earth Planet. Sci. Lett., vol. 208, 69–84.

Lowenstein, T. K., Timofeeff, M. N., Brennan, S. T., Hardie, L. A., and Demicco, R. V., 2001, Oscillations in Phanerozoic seawater chemistry: Evidence from fluid inclusions. Science, vol. 294, 1086–1088. http://doi .org/10.1126/science.1064280

McArthur, J. M., Howarth, R. J., and Bailey, T. R., 2001, Strontium isotope stratigraphy: LOWESS version 3. Best-fit line to the marine Sr isotope curve for 0 to 509 Ma and accompanying look-up table for deriving numerical age. J. Geol., vol. 109, 155–169.

McArthur, J. M., Howarth, R. J., and Shields, G. A., 2012, Strontium isotope stratigraphy. In Gradstein, F. M., Ogg, J. G., Schmitz, M., and Ogg, G. (eds), The Geologic Time Scale 2012. Elsevier, pp. 127–144 (Comprehensive Phanerozoic seawater Sr curve showing selected data that define the curve. Update in 2020 listed in references.)

McArthur, J. M., Howarth, R. J., Shields, G. A., and Zhou, Y., 2020, Strontium isotope stratigraphy. In Gradstein, F. M., Ogg, J. G., Schmitz, M. D., and Ogg, G. M. (eds), The Geologic Time Scale, vol. 1. Elsevier, pp. 211–238.

McCauley, S., and DePaolo, D. J., 1997, The marine ^{87}Sr/^{86}Sr and δ^{18}O records, Himalayan alkalinity fluxes and δ. In Ruddiman, W. F. (ed), Tectonic Uplift and Climate Change. Plenum, pp. 427–467.

Mokadem, F., Parkinson, I. J., Hathorne, E. C. et al., 2015, High precision radiogenic strontium isotope measurements of the modern and glacial ocean: Limits on glacial-interglacial variations in continental weathering. Earth Planet. Sci. Lett., vol. 415, 111–120.

Müller, M. N., Krabbenhoft, A., Vollstaedt, H., Brandini, F. P., and Eisennhauer, A., 2018, Stable isotope fractionation of strontium in coccolithophore calcite: Influence of temperature and carbonate chemistry. Geobiology, vol. 16, 297–306.

Muller, D. W., and Mueller, P. A., 1991, Origin and age of the Mediterranean evaporates: Implications from Sr isotopes. Earth Planet. Sci. Lett., vol. 107, 1–12.

Palmer, M. R., and Edmond, J. M., 1989, The strontium isotope budget of the modern ocean. Earth Planet. Sci. Lett., vol. 92, 11–26.

Park, Y., Swanson-Hysell, N. L., MacLennan, S. A., et al., 2020, The lead-up to the Sturtian Snowball Earth: Neoproterozoic chemostratigraphy time-calibrated by the Tambien Group of Ethiopia. Geol. Soc. Am. Bull. Vol. 132, 1119–1149.

Peucker-Ehrenbrink, B., and Fiske, G. J., 2019, A continental perspective of the seawater $^{87}Sr/^{86}Sr$ record: A review. Chem. Geol., vol. 510, 140–165.

Raymo, M. E., and Ruddiman, W. F., 1992, Tectonic forcing of late Cenozoic climate. Nature, vol. 359, 117–122.

Raymo, M. E., Ruddiman, W. F., and Froelich, P. N., 1988, Influence of late Cenozoic mountain building on ocean geochemical cycles. Geology, vol. 16, 649–653.

Richter, F.M. and D.J. DePaolo, 1988, Diagenesis and Sr isotopic evolution of seawater using data from DSDP 590B and 575: Earth Planet. Sci. Lett. vol. 90, 382–394.

Saltzman, M. R., Edwards, C. T., Leslie, S. A. et al., 2014, Calibration of a conodont apatite-based Ordovician $^{87}Sr/^{86}Sr$ curve to biostratigraphy and geochronology: Implications for stratigraphic resolution. Geol. Soc. Am. Bull., vol. 126, 1551–1568. http://doi.org/10.1130/B31038.1

Sedlacek, A. R., Saltzman, M. R., Algeo, T. J. et al., 2014, $^{87}Sr/^{86}Sr$ stratigraphy from the early Triassic of Zal, Iran: Linking temperature to weathering rates and the tempo of ecosystem recovery. Geology, vol. 42, 779–782.

Shields, G., and Veizer, J., 2002, Precambrian marine carbonate isotope database: Version 1.1. Geochem. Geophys. Geosyst., vol. 3. http://doi.org/10.1029/2001GC000266

Sinnesael, M., Montanari, A., Frontalini, F. et al., 2019, Multiproxy Cretaceous-Paleogene boundary event stratigraphy: An Umbria-Marche basinwide perspective. In Koeberl, C., and Bice, D. M. (eds), 250 million years of Earth history in central Italy: Celebrating 25 years of the Geological Observatory of Coldigioco, Special Paper 542. Geological Society of America, pp.133–158. http://doi.org/10.1130/2019.2542(07)

Song, H., Wignall, P. B., Tong, J. et al., 2015, Integrated Sr isotope variations and global environmental changes through the Late Permian to early Late Triassic. Earth Planet. Sci. Lett., vol. 424, 140–147.

Steuber, T., and Veizer, J., 2002, Phanerozoic record of plate tectonic control of seawater chemistry and carbonate sedimentation. Geology, vol. 30, 1123–1126.

Swanson-Hysell, N. L., and Macdonald, F. A., 2017, Tropical weathering of the Taconic orogeny as a driver for Ordovician cooling. Geology, vol. 45, 719–722. http://doi.org/10.1130/G38985.1

Turchyn, A. V., and DePaolo, D. J., 2019, Seawater chemistry through Phanerozoic time. Annu. Rev. Earth Planet. Sci., vol. 47, 197–224.

Veizer, J., 1989, Strontium isotopes in seawater through time. Annu. Rev. Earth Planet. Sci., vol. 17, 141–167. http://doi.org/10.1146/annurev.ea.17.050189.001041

Veizer, J., and Compston, W., 1974, $^{87}Sr/^{86}Sr$ composition of seawater during the Phanerozoic. Geochim. Cosmochim. Acta, vol. 38, 1461–1484.

Zhang, S. and DePaolo, D.J., 2020, Equilibrium calcite fluid Sr/Ca partition coefficient from marine sediment and pore fluids. Geochim. Cosmochim. Acta, vol. 289, 33–46.

Zuza, A. V., and Yin, A., 2017, Balkatach hypothesis: A new model for the evolution of the Pacific, Tethyan, and Paleo-Asian oceanic domains. Geosphere, vol. 13, 1664–1712. http://doi.org/10.1130/GES01463.1

Acknowledgments

The authors acknowledge support from the US National Science Foundation, the Petroleum Research Fund, and the University of California over many years that made their research on this topic possible. The manuscript benefited from the comments and suggestions of two anonymous reviewers.

Cambridge Elements ☰

Geochemical Tracers in Earth System Science

Timothy Lyons
University of California

Timothy Lyons is a Distinguished Professor of Biogeochemistry in the Department of Earth Sciences at the University of California, Riverside. He is an expert in the use of geochemical tracers for applications in astrobiology, geobiology, and Earth history. Professor Lyons leads the "Alternative Earths" team of the NASA Astrobiology Institute and the Alternative Earths Astrobiology Center at University of California, Riverside.

Alexandra Turchyn
University of Cambridge

Alexandra Turchyn is a University Reader in Biogeochemistry in the Department of Earth Sciences at the University of Cambridge. Her primary research interests are in isotope geochemistry and the application of geochemistry to interrogate modern and past environments.

Chris Reinhard
Georgia Institute of Technology

Chris Reinhard is an Assistant Professor in the Department of Earth and Atmospheric Sciences at the Georgia Institute of Technology. His research focuses on biogeochemistry and paleo-climatology, and he is an Institutional PI on the "Alternative Earths" team of the NASA Astrobiology Institute.

About the Series

This innovative series provides authoritative, concise overviews of the many novel isotope and elemental systems that can be used as "proxies" or "geochemical tracers" to reconstruct past environments over thousands to millions to billions of years – from the evolving chemistry of the atmosphere and oceans to their cause-and-effect relationships with life.

Covering a wide variety of geochemical tracers, the series reviews each method in terms of the geochemical underpinnings, the promises and pitfalls, and the "state-of-the-art" and future prospects, providing a dynamic reference resource for graduate students, researchers, and scientists in geochemistry, astrobiology, paleontology, paleoceanography, and paleoclimatology.

The short, timely, broadly accessible papers provide much-needed primers for a wide audience – highlighting the cutting edge of both new and established proxies as applied to diverse questions about Earth system evolution over wide-ranging time scales.

Cambridge Elements ☰

Geochemical Tracers in Earth System Science

Printed in the United States
by Baker & Taylor Publisher Services

Printed in the United States
by Baker & Taylor Publisher Services